Bibliographic information published by the German National Library:

The German National Library lists this publication in the National Bibliography; detailed bibliographic data are available on the Internet at http://dnb.dnb.de .

Imprint:

Print and binding: Books on Demand GmbH, Norderstedt Germany
ISBN: 9783346139849

This book at GRIN:

https://www.grin.com/document/516601

Louie Sanlad

Air Pollution Awareness in the Philippines. Its Practical Measure for Prevention

Clean Air Act, Causes and Effects of Air Pollution, and Measures for Prevention

GRIN Verlag

GRIN - Your knowledge has value

AIR POLLUTION AWARENESS OF PRE-SERVICE TEACHERS IN THE PHILIPPINES

Rationale of the Research

The threat of environmental pollution has become one of the chief concerns of this decade and perhaps the rest of this century. Although many individuals think the industrialists are the villains who pollute the environment, the major villains are the consumers who desire and demand new, more, faster, and bigger playthings without thinking about the cost to the environment. Though many living things emit carbon dioxide when they breathe, the gas is widely considered to be a pollutant when associated with cars, planes, power plants, and other human activities that involve the burning of fossil fuels such as gasoline and natural gas. That is because carbon dioxide is the most common of the greenhouse gases, which trap heat in the atmosphere and contribute to climate change (Delgado, A. 2019).

Alongside being an increasingly important problem, air pollution is a major issue regarding various fields such as environment, health, economy and politics. Pollution is now a common place term, which our ears are attuned to. We hear about the various forms of pollution and read about it through the mass media. Air pollution is one such form that refers to the contamination of the air, irrespective of indoors or outside. A physical, biological or chemical alteration to the air in the atmosphere can be termed as pollution. It occurs when any harmful gases, dust, smoke enters into the atmosphere and makes it difficult for plants, animals and humans to survive as the air becomes dirty. It is a mix of particles and gases that can reach harmful concentrations both outside and indoors. Its effects can range from higher disease risks to rising temperatures. Soot, smoke, mold, pollen, methane, and carbon dioxide are a just few examples of common pollutants. Combining these definitions, air pollution in general can be described as the increase in the rate of harmful gases and particles in the atmosphere (Özey, 2009).

The pervasive social and environmental impacts of air pollution make it an important marker for sustainable development across all levels of economic development. The drivers behind pollution differ by economic structure, however, and developing countries have different problems than developed countries. Air pollution is an important indicator for environmental quality and public health in developing regions, as economic expansion contributes to higher pollution levels (World Bank and Institute for Health Metrics and Evaluation, 2016).

Philippines as a developing country, it deals with many socio-economic and developmental issues. In the face of these issues, some are concerned that over the past two years and in spite of the current administration's promises, environmental protection appears to have taken a backseat. The government, of course, is responsible for creating and enforcing policies and laws so that its citizens may breathe clean air, as well as enjoy all natural resources based on the principles of sustainable development. In the southeast Asian region — where the Philippines is located — the average PM2.5 concentration is 21 µg/m3 annually, over twice the recommended value (Tomacruz, S., 2018).

Moreover, the Philippines' annual PM2.5 concentration rests at 18.4 µg/m3 according to the 2016 WHO report. This value is approximately 80%

higher than the safe levels indicated by WHO. Aside from national data, WHO also collected data from the major cities of participating countries. Meanwhile, Cebu, Dagupan, and Manila also failed, registering values of 54, 51, and 55 µg/m3, respectively (Ambag, R. 2018),

Air quality has historically been poor in the Philippines. After the passing of the Clean Air Act in 1999, the Philippine area struggled to meet the Environmental Protection Act. Improving the educational level and raising environmental awareness are essential for reducing the air pollution (Selden & Song 1994). From early years on the importance of Environmental Education (EE) arose and teachers are the key point of EE.

Conceptual Background

This research is anchored on Republic Act No. 8749, also known as Philippine Clean Air Act of 1999. It is a comprehensive air quality management policy and program which aims to achieve and maintain healthy air for all Filipinos. The Philippine Clean Air Act of 1999 was signed into law by former President Joseph Estrada in July 27, 1999. It is an act providing for a comprehensive air pollution control policy and for other purposes was landmark legislation in Philippine environmental protection. The law has 56 Sections and divided into seven (7) chapters namely, General Provisions; Air Quality Management System; Fuels, Additives, Substances and Pollutants; Institutional Mechanism, Actions; Fines and Penalties; and Final Provisions.

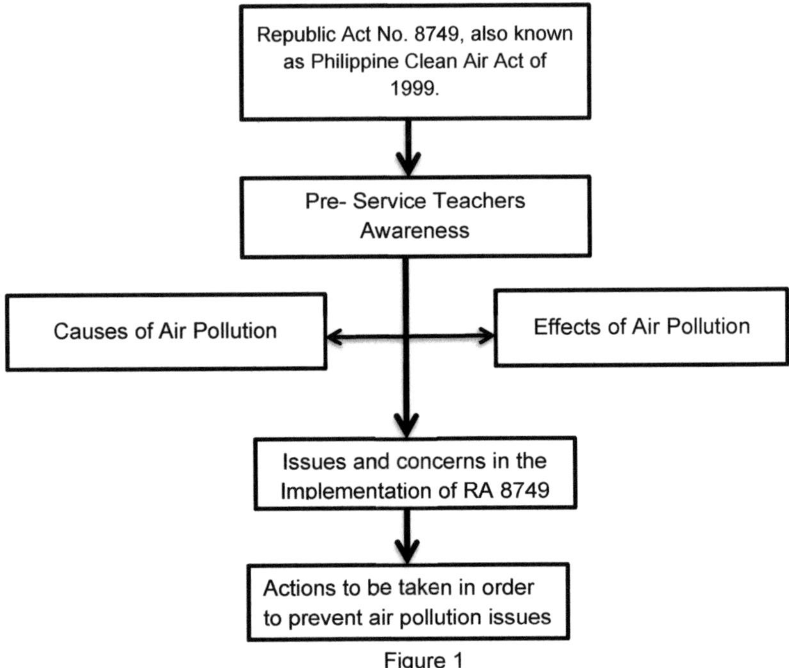

Figure 1

Conceptual framework of the Research

The principle behind the creation of the law is where the state recognizes its responsibility in protecting the rights of people in living in a balanced ecology where the quality of air is adequate. The state also acknowledges that social-justice is primary concern when this principal right is violated by developing mechanisms for preventing, managing, restricting, and penalizing air pollution. Importantly, the recognition of these rights were stipulated in Section 4 which the state shall seek to guarantee the following enjoyment of these rights. The right to breathe clean air is one of those rights. The fact that breathing clean air has to be a fundamental right tells us that the society we have built has undermined the importance of ecological balance in living a sustainable life. It is frowning that reality has to be as grim as the laws that are written to promote quality of life.

Technical definitions of scientific concepts in environmental science were included in the legislation as operative terms used in the act. One of the terms that are important to note for future environmental planners is on Chapter 1, Article 2, Section 5, Item g). Eco-profile – means the geographical –based instrument for planners and decision-makers who present an evaluation of the environmental quality and carrying capacity of an area. It is the result of the integration of primary and secondary data and information on natural resources and anthropogenic activities on the land which are evaluated by various environmental risk assessment and forecasting methodologies that enable the Department to anticipate the type of development control necessary in the planning area. The definition beforehand speaks true to the importance of the planning discipline in the assurance of sustainable development with regards to the Philippine Clean Air Act.

Gathering the data needed to plan for the effects of air pollution in human settlements was recognized in a pivotal role for Air Quality Monitoring and Information Network. Reporting the findings of these monitoring activities and research guides the decision maker and planner in making decisions that would affect the quality of life of the population. Implementing this under the Integrated Air Quality Improvement Framework would be challenging, since it seeks to prescribe the emission reduction goals using permissible standards, control strategies and control measures to be undertaken within a specified time period, including cost-effective use of economic incentives, management strategies, collective action, and environmental education and information. The monitoring and information network seeks to understand and implement the air quality standards based not only on the World Health Organization Standards but also not less as stringent than other internationally accepted standards.

The air quality standard followed by the Environmental Management Bureau which was assigned by law to take on the responsibility of air quality management was defined under Section 12 or the Ambient Air Quality Guideline Values and Standards. The standards are well articulated in following international safety standards of acceptable particulate matter overtime from different pollutants.

Quality control follows into the realm of burning garbage or incineration. Section 20 of the law clearly states a ban on incineration that defined the municipal, bio0medical and hazardous wastes burning. Yet traditional methods of burning or siga are still allowed, as is with kaingin as a traditional agricultural practice. This part of the law does not seem to jive well with internationally recognized practices that also ban small pit fire burning or bush burning. This part of the law showcases the weakness of policy in pandering to the harmful practices of past just to make way for public acceptance.

In article 4 of chapter 2 of this act gave the responsibility and jurisdiction of implementing air quality controls to the Department of Transportation and Communication on monitoring and processing of permits for emission standards of pollution from motor vehicles. Any observer in EDSA could easily conclude that there is a failure of implementation in this regard. Public transport vehicles are number one in terms of violators of this law due to the alleged collusion of private testing centers in passing unworthy vehicles for a fixer's fee. The standards are there, but as with any bureaucracy, implementation of rules and regulations is a different matter.

The law is a sound piece of legislation although it is weak in institutional support. The EMB or Environmental Management Bureau is a small office that monitors not only air quality but water and other resources as well. With a bureau sized staff, operations for implementing the law is inefficient. The resolution for this good law is to upgrade the Bureau to a separate Department of Environmental Protection with its own Cabinet level secretary to oversee its functions and mandate.

We depend on air for our respiratory needs. So, air pollution causes injury to all living organisms. In case of plants, the growth and yield of crops are reduced and cause premature death. In animals including man, serious metabolic and respiratory diseases are manifested due to air pollution. Air pollution is also called as atmospheric pollution. The atmosphere is an invisible layer of gases that surround the earth. Sources of Air Pollution:

Air pollution is a concern which poses threat to human health and likewise greatly impacts the ecosystem and the environment. An expert on air quality and pollution, Environment and Management Bureau Consultant Engr. Reynaldo Tejada shares his knowledge on the impact of air quality and pollution in human health, the ecosystem and the environment in a presentation held here recently on air modeling in the Baguio-La Trinidad-Itogon-Sablan-Tuba-Tublay areas. Even with the passage of the Clean Air Act which puts in place air pollution control and policy, Tejada said there are reasons why one should be concerned in air pollution as it poses significant threat to the human health and atmosphere (Susan C, Aro, 2018). Its effect in the human body leads to various health-related diseases such as respiratory disease, decreased lung functions, cancers, eye and throat irritation, he adds. At high risk are children and elderly and those individuals with asthma and cardiopulmonary diseases. As to the

According to The Atlantic, the guidelines, proposed in April, appear to target regulations that limit particulate air pollution. They are backed by a 1993 study showing that Americans living in cities with higher particulate levels suffered health consequences. That study was the "biggest science fraud that has gone on in this country's history," Steven Milloy of the conservative Heartland Institute tells The Atlantic. Other backers of the rule include members of industry trade groups, the Associated Press reports, but critics outnumbered supporters at yesterday's hearing. The proposal has also drawn opposition from the editors of scientific journals and dozens of public health, medical, academic, and scientific groups (Williams H., 2018).

Around seven million people die annually from exposure to fine particles in polluted air that penetrate deep into the lungs and cardiovascular system, thus causing stroke, heart disease, lung cancer, chronic obstructive pulmonary diseases and respiratory infections, including pneumonia. Air pollution is said to be a critical risk factor for noncommunicable diseases (NCDs), causing an estimated one-quarter (24 percent) of all adult deaths from heart disease, 25 percent from stroke, 43 percent from chronic obstructive pulmonary disease and 29 percent from lung cancer (Robles, Nathalie, 2018). Major sources of air pollution include the inefficient use of energy by households, industry, the agriculture and transport sectors, and coal-fired power plants. In some regions, sand and desert dust, waste burning and deforestation are additional sources of air pollution. Air quality can also be influenced by natural elements such as geographic, meteorological and seasonal factors.

The Philippines' Clean Air Act of 1999 (Republic Act No. 8749) is almost two decades old, yet enforcement seems to be an issue as the quality of air in the Philippines continues to deteriorate. The Department of Environment and Natural Resources (DENR) says it is working with the Department of Health and WHO to decrease the environmental health risk of air pollution by improving the linkage of air quality and health monitoring, reviewing the air quality index, and strengthening the health impact assessment for projects and activities that may cause air pollution. The DENR's Air Quality Management Bureau and the Land Transportation Office (LTO) are working together to track and apprehend smoke belchers. Smoke belchers are subject to fines, license suspensions or franchise cancellations (Jimeno . Karen V., 2018).

Decades after the passage of the Clean Air Act, the fight to improve the quality of air for Filipinos has little progressed. This was seen as a major milestone in the implementation of the law. After this, however, things came to a sort of slide. The argument is that the quality of air in Metro Manila, being the key measurement of nationwide air pollution, is improving. But is definitely not anywhere nearer what is ideally regarded as clean air. Today, Metro Manila continues to experience concentrations of particulate matter that routinely exceed acceptable levels. Carbon monoxide and nitrogen oxide levels have also been monitored to occasionally exceed standards.

Sadly, what was hailed as a landmark piece of legislation to protect the environment seems to have lost its fighting fizz. Implementing the law, apparently, is more than just calling for a ban on lead or on mass incineration. Despite all the funds that have been funneled to assist our local executives in enforcing the Clean Air Act, there seemed to be little motivation for the respective government agencies to pursue their mandates. Roadside smoke emission tests, for one, are now being conducted with even less frequency. There is talk that the once-a-year mandatory smoke emission test, a requirement for getting a permit to operate a motor vehicle, can now be bought. In fact, you don't need to bring your car to an accredited testing center if you know who to talk to.

The Philippines is currently pushing back against air pollution through the implementation of various laws and ordinances. One of these is Republic Act No. 8749, or the Philippine Clean Air Act of 1999. RA 8749 currently serves as the foundation for the country's air quality management. It aims to raise awareness about pollution prevention through programs such as Linis/Ligtas Hangin, coupled with Bantay Tambutso, Bantay Tsimnea, and Bantay Sunog. The RA also aims to enforce regulatory standards upon stationary sources of pollution, such as factories and power plants. RA 8749 encompasses the cooperation of agencies such as the Department of Environment and Natural Resources (DENR), the Department of Transportation and Communication (DOTC), the Department of Science and Technology (DOST), the Department of Trade and Industry (DTI), the Department of Energy (DOE), and many others (Ambag, R. 2018).

Meanwhile, WHO Representative to the Philippines Dr. Gundo Weller stated:

> *"Air pollution continues to pose a dangerous threat to the health of the Filipinos, especially those living in urbanized communities...With the health and environment sectors, we are working together to reduce the harmful impact of these environmental risks to people's health and well-being."*

Air Pollution in the Philippines

Air pollution occurs when gases, dust particles, fumes (or smoke) or odor are introduced into the atmosphere in a way that makes it harmful to humans, animals and plant. This is because the air becomes dirty (contaminated or unclean). Anything additional gas, particles or odors that are introduced into the air (either by nature or human activity) to destroy this natural balance can be called air pollution. It has been present in many civilizations from the very time that people started to make fires. It is the presence of harmful materials in earth's atmosphere. Our atmosphere is gaseous system that supports life in our planet. It is a very important factor that makes life available on earth. It is a threat to the many aspects of life.

MAJOR CAUSES: POLLUTANTS

Pollutants are the so called causes to air pollution which is comprised of different chemicals and toxic wastes. Major causes pollutants are Sulfur Oxide, Nitrogen Oxide , Carbon , Monoxide , Volatile Organic Compounds, Particulates, Toxic Metals, Chlorofluorocarbons, Ammonia and odor (Libiternos, Mapi, 2014).

Sulfur Oxide is a colorless gas that cannot be seen. Its most common natural source is volcanoes and the most common man-made source is the burning of fuels like coal, oil, gas, petrol and diesel.

Nitrogen Oxide is ac reddish, brown gas that is poisonous and very harmful to human beings. It would make breathing difficult, give chest pains and might damage one's lungs once inhaled.

Carbon Monoxide is a gas that cannot be seen nor smelled. It is also poisonous and breathing a lot of it in can cause death because our lungs confuse it with oxygen and put it into our blood by mistake.

Volatile Organic Compounds is primary precursors to the formation of ground level ozone and particulate matter in the atmosphere which are the main ingredients of the air pollutant referred to as smog.

Particulates is an unseen things that are floating around in the air. It comes in almost any shape or size, and can be solid particles or liquid droplets. It is divided into two major groups. These groups differ in many ways. One of the differences is size, we call the bigger particles PM10 and we call the smaller particles PM2.5.

Toxic Metals refers to any metallic chemical element that has a relatively high density and is toxic or poisonous at low concentrations.

Chlorofluorocarbons is widely used as coolants in refrigeration and air conditioners, as solvents in cleaners, particularly for electronic circuit boards, as a blowing agents in the production of foam (for example fire extinguishers), and as propellants in aerosols.

Ammonia is a common naturally occurring substance. It is also manufactured by man. At normal environmental conditions, pure ammonia is a colorless, pungent-smelling, caustic (corrosive) gas. It is stored under high pressure as a liquid. It is highly soluble in water and reacts with acids to form ammonium salts. The main local problem of ammonia released into air is the unpleasant odor, which is detectable even at low concentrations.

Odors is caused by one or more volatilized chemical compounds, generally at a very low concentration, that humans or other animals perceive by the sense of olfaction.

Human activities that create and emit these elements that stated earlier are industrial processes,, electric discharge, combustion, aerosols, agricultural processes, garbage, sewage and nuclear explosions.

Industrial processes are procedures involving chemical or mechanical steps to aid in the manufacture of an item or items, usually carried out on a very large scale. Industrial processes are the key components of heavy industry.

Electric discharge describes any flow of electric charge through a gas, liquid or solid.

Combustion (burning) is the sequence of exothermic chemical reactions between a fuel and an oxidant accompanied by the production of heat and conversion of chemical species. The release of heat can produce light in the form of either glowing or a flame.

Aerosols - is a colloid of fine solid particles or liquid droplets, in air or another gas.

Agricultural processes are specific methods which, when applied to agriculture, create food for consumers.

Garbage- wasted or spoiled food and other refuse, as from a kitchen or household.

Sewage is a waste matter from domestic or industrial establishments that is carried away in sewers or drains.

Nuclear Explosions is an explosion that occurs as a result of the rapid release of energy from a high-speed nuclear reaction

NATURAL SOURCES OF AIR POLLUTION

Some of the natural sources of air pollution are organic compounds from plants, sea salt, suspended soils and dusts. Other natural sources are released during catastrophes such as volcanic eruptions and forest fires. Large amounts of harmful gases and smoke are released which can increase background pollution levels for years - even in areas far away from the original source. Ozone is one of the most common n atural air pollutants. Libiternos, Mapi (2014) enumerates the natural sources of air pollution in the Philippines as dust, Methane, Radon, smoke vegetation and volcanic activities.

Dust consists of particles in the atmosphere that comes from various sources such as soil, dust lifted by weather (an aeolian process), volcanic eruptions, and pollution.

Methane is a chemical compound with the chemical formula CH_4 (one atom of carbon and four atoms of hydrogen). It is the simplest alkane and the main component of natural gas.

Radon is a known pollutant emitted from geothermal power stations because it is present in the material pumped from deep underground.

However, it disperses rapidly, and no radiological hazard has been demonstrated in various investigations.

Smoke is a collection of airborne solid and liquid particulates and gases.

Vegetation emits environmentally significant amounts of VOCs on warmer days. These VOCs react with primary anthropogenic pollutants—specifically, NOx, SO2, and anthropogenic organic carbon compounds—to produce a seasonal haze of secondary pollutants.

Volcanic Activities produces sulfur, chlorine, and ash particulates.

EFFECTS OF AIR POLLUTION

It is impossible to describe the whole extent of potential and actual damage caused by all forms of air pollution. The main consequences of air air pollution are on the environment and human's health.

ENVIRONMENTAL IMPACTS

Environmental degradation has increased over the last several decades. Vehicular emissions, rapid industrialization and urbanization are the main sources of air pollution Singh, A., Gupta, V. and Sasm S. (2009) categorized the environmental effects of air pollution as global warming, formation of photochemical smog, formation of acid rain, aerosol formation, and depletion of ozone.

Global warming. According to estimates, at the current rate of increase, the average global temperature up by 3°C to 8°C in the next 100 years. This will affect the climate of different regions, distribution of plants and animals, disturbance in agriculture and food production, melting of snow caps and resultant increase in sea levels. This will submerge parts of coastal cities of Calcutta, New York, London and other major cities.

Formation of Photochemical Smog. When pollutants like hydrocarbons and nitrogen oxides combine in the presence of sunlight, smog is formed. This is a mixture of gases and since it is formed by photochemical reactions, it is called the photochemical smog. The word 'smog' is derived from the two words-smoke and fog. It forms a yellowish brown haze especially during winter and hampers visibility. It also causes many respiratory disorders and allergies as it contains polluting gases.

Formation of Acid Rain. Sulphur dioxide and nitrogen oxides react with water in the atmosphere producing sulphuric acid and nitric acid. These acids come down along with the rain. This phenomenon is called acid rain. The pH of acid rain varies from 3-6. The composition of acid rain is sulphuric acid, nitric acid and weak carbonic acid. Its adverse affects on the environment include: causes respiratory and skin disorders, affects productivity of plants by damaging the leaves, enters the soil and affects the soil, pH and causes leaching, enters the ground and river waters which causes harm to the aquatic life, causes damage to marble and thus damages buildings and monuments (like Taj Mahal).

Aerosol Formation. Aerosol is formed by the dispersion of solid or liquid matter in the atmosphere. If the aerosols form a thick layer in the troposphere, they affect the weather conditions by blocking the solar radiation. Aerosols are also deposited on the leaves and affect the photosynthesis. Aerosols disperse the organic metallic pollutants far and wide.

Depletion of Ozone. The stratosphere of the atmosphere has ozone (O_3). Ozone is known to absorb the Ultraviolet (UV) rays present in the sun's radiation and protects us from the harmful effects of the UV rays. However, hydrocarbons such as the chlorofluorocarbons (CFCs) destroy the ozone molecules which deplete the ozone layer. Ozone holes have been detected in the atmosphere which permits the UV rays to reach the earth's surface. The harmful effects of the UV rays are visible in the countries such as Australia and New Zealand where the rate of skin cancer is higher than the other regions of the world.

EFFECTS OF AIR POLLUTION TO HUMAN'S HEALTH

In terms of health hazards, every unusual suspended material in the air, which causes difficulties in normal function of the human organs, is defined as air toxicants. According to available data, the main toxic effects of exposure to air pollutants are mainly on the respiratory, cardiovascular, ophthalmologic, dermatologic, neuropsychiatric, hematologic, immunologic, and reproductive systems. However, the molecular and cell toxicity may also induce a variety of cancers in the long term (Otsuki, Nakano, 2013). On the other hand, even small amount of air toxicants is shown to be dangerous for susceptible groups including children and elderly people as well as patients suffering from respiratory And Cardiovascular Diseases.

Respiratory Disorders. Because most of the pollutants enter the body through the airways, the respiratory system is in the first line of battle in the onset and progression of diseases resulted from air pollutants. Depending on the dose of inhaled pollutants, and deposition in target cells, they cause a different level of damages in the respiratory system. In the upper respiratory tract, the first effect is irritation, especially in trachea which induces voice disturbances. Air pollution is also considered as the major environmental risk factor for some respiratory diseases such as asthma and lung cancer. Air pollutants, especially PMs and other respirable chemicals such as dust, O3, and benzene cause serious damage to the respiratory tract. Asthma is a respiratory disease which may be developed as a result of exposure to air toxicants (Valavanidis A, Vlachogianni T, Fiotakis K, Loridas S., 2013).

Cardiovascular Dysfunctions. Many experimental and epidemiologic studies have shown the direct association of air pollutant exposure and cardiac-related illnesses. Air pollution is also associated with changes in white blood cell counts which also may affect the cardiovascular functions. On the other hand, a study on animal models suggested the close relationship between hypertension and air pollution exposure. The traffic-related air pollution, especially exposure to high levels of NO2, is associated with right and left ventricular hypertrophy. In addition to the antidote therapy that exists only for a few cardiotoxic substances like CO, usual treatment of cardiovascular diseases should be carried out (Nogueira JB., 2009).

Neuropsychiatric Complications. The relationship between exposure to air suspended toxic materials and nerve system has always been argued. However, it is now believed that these toxic substances have damaging effects on the nervous system. The toxic effect of air pollutants on nerve system includes neurological complications and psychiatric disorders. Neurological impairment may cause devastating consequences, especially in infants. In contrast, psychiatric disorders will induce aggression and antisocial behaviors. Recent studies have reported the relationship between air pollution and neurobehavioral hyperactivity, criminal activity, and age-inappropriate behaviors (Newman NC, Ryan P, Lemasters G, Levin L, Bernstein D, Hershey GK, et al., 2013)

PRACTICAL MEASURES TO REDUCE AIR POLLUTION

The industrialization of societies is necessary to develop, but a long-term health problem and ecological impacts of such growth should always be considered prior to imposing a large financial burden on the societies. Therefore, it is suggested to adopt a balance between economic development and air pollution by legislating policies to control all activities resulting in air pollution (Dincer I., 2000). There are some temporary but not trustworthy strategies to reduce air pollution in Iran. For example, increasing the price of fuels, planting trees around and inside the city, replacing old cars with modern ones, and increasing road taxes and car insurance may reduce the amount of air pollutants, but in order to keep the constancy or even optimization, these strategies should be continued. Therefore, urgent and concerted actions at national and international levels are required.

Standardization of vehicle's fuel as much as possible and also finding a new source of energy for motor engines has attracted great attention. A great part of emission comes from vehicle exhaust, especially those which use diesel and gasoline. Using other clean source of energy such as compressed natural gas (CNG), liquefied natural gas (LNG), and alcohol is of great interests. Hence, exhorting researchers and also companies in the era of interest to find a way for replacing petrol and other fossil fuels with new suitable power generation sources will be beneficial. Expanded of more CNG and LNG stations in big cities of Iran and producing more vehicles using CNG and LNG are recommended (Goldemberg J, Johansson TB, Reddy AK, Williams RH, 2001).

Standardization of motor engines and manufacturing engines with low fuel consumption. It is another strategy to reduce the level of air pollutants. Surely, motor vehicles will not use fossil fuel and derivatives anymore in the near future. Recently, some car makers in Japan and Western countries have made electrical cars that use electricity from a battery storage for low speeds which is usually enough for driving inside the cities with traffic and/or controlled speeds. Hence, designing new motor engines, companies of interests should look forward (Faiz A, Weaver CS, Walsh MP., 1996).

Improving public transportation systems by using more subways (metro), trams, and electrical bus routes. Reducing the costs for the people who are using such systems is an optimal solution for lowering air pollution. The people and governments will gain profits from reducing air contaminations in the long term, both economically and in cases of health issues. It is recommended to expand the metro lines in these cities (Wong EY, Gohlke J, Griffith WC, Farrow S, Faustman EM., 2004).

Increasing the cost of fuel in Iran can be considered as an effective solution to reduce the proportion of air contaminants. According to the report by Barnett and Knibbs, higher fuel price is associated with lower air pollution level (Barnett AG, Knibbs LD. , 2014).

Imposing penalties for polluting industries and implementing low tax policy for clean technologies. Applying more taxes on automobiles in Iran, particularly on those older than 20 years to distinguish between dirty and clean vehicles. The government should establish exhorting plans for car makers and other producers who adhere to environmental standards. Moreover, exhorting plans should be designed for all sources of emissions. These plans may contain discount on taxes or other financial supports for customers and producers (Becker R, Henderson V., 2000).

Communication and Collaboration. Since the phenomenon of air pollution was a global issue over the centuries since the industrial revolution, it is proposed to establish an interdisciplinary academic field on air pollution. It is also suggested that more communication and collaboration between specialists in different sciences including toxicology, environmental health, analytical chemistry, mechanics, and applied physics will be performed

Monitoring. Continuous monitoring of air quality, designing and developing tools to identify the pollutants, finding the origin of the particles, and the use of particulate filter for diesel engines and other nonroad cars are other suggested practical approaches to reduce air pollution (Dincer I., 2000).

Extensive media campaign to increase public awareness about air quality, environmental, and public health issues (Holman C., 1999).

AIR POLLUTION PREVENTION

Solution efforts on pollution are always a big problem. This is why prevention interventions are always a better way of controlling air pollution. These prevention methods can either come from government (laws) or by individual actions. In many big cities, monitoring equipment has been installed at many points in the city. Authorities read them regularly to check the quality of air. Let's see more below:

Government (or Community) Level Prevention.

Governments throughout the world have already taken action against air pollution by introducing green energy. Some governments are investing in wind energy and solar energy, as well as other renewable energy, to minimize burning the of fossil fuels, which cause heavy air pollution.

Governments are also forcing companies to be more responsible in their manufacturing activities, so that even though they still cause pollution, they are a lot controlled.

Car manufacturing companies are also building more energy efficient cars, which pollute less than before.

Individual Level Prevention

Encourage your family to use the bus, train or bike when commuting. If we all do this, there will be fewer cars on the road and less fumes produced.

Use energy (light, water, boiler, kettle and fire woods) wisely. This is because lots of fossil fuels are burned to generate electricity, and so if we can cut down the use, we will also cut down on the amount of pollution we create.

Recycle and re-use things. This will minimize the dependence of producing new things. Remember manufacturing industries create a lot of pollution, so if we can re-use things like shopping plastic bags, clothing, paper and bottles, it can help.

In the study of Olufemi C. Adejoke,Andile Mji & Murembiwa S. Mukhola, (2017) on Students' and Teachers' Awareness of and Attitude towards Environmental Pollution determined the participants' biographical factors that had an effect on environmental Awareness and Attitude. For teachers, on the other hand, no significant effects were established for all the biographical variables tested. More pleasing was that teachers had higher scores with respect to environmental Awareness and Attitude compared with their students. The researchers argue that these findings are important because it is the teachers who will hopefully pass the proverbial baton to their students, for them to embrace environmental issues in the air polluted area whence they reside.

When learning about the environment's declining health it is easy to feel discouraged, but what keeps us fighting for a healthy world is the future of our children. They should not have to inherit our environmental problems and in order to keep their future bright, spreading awareness is imperative. Aygun, M., Ceylan, D. & Gungoren, S., (2017) in analysis of their study on Air Pollution Awareness in the Scope of the Community Service Practices Course indicates that learning environments need to be designed that focus on fundamental concepts, causes-effects, future situation and precautions in order to increase air pollution awareness. To achieve an interdisciplinary approach, a two stage learning environment has been designed by relating these focus points with Turkish, Social Studies, Science, Mathematics and Public Administration disciplines respectively. These two stages are preparing the discipline-based course plans and materials; and carrying out the interdisciplinary plans with students by pre-service teachers.

Significance of the Study

Government sector. This will help to strictly monitor the policy that protects and advances the right of the people to a balanced and healthful ecology in accord with the rhythm and harmony of nature and to encourage the participation of an informed and active public in air quality planning and monitoring.

Department of Environment and Natural Resources. This will serve as the official blueprint with which all government agencies must comply with to attain and maintain ambient air quality standards.

Local Government Unit. This will help develop a procedure for carrying out the action plan for their jurisdiction and maintain its authority to independently inspect the enforcement procedure to closely supervise all or parts of the air quality action plan until such time the concerned can assume the function to enforce the standards set.

Education sector. This will help formulate program or activity that shall include the setting up of a funding or guarantee mechanism for clean-up and environmental rehabilitation and compensation for personal damages.

Researchers. This research will help the researchers to explore the possibility of setting up instructional and precautionary programs for their learners and fellow teachers which guarantee their benefits from the awareness of air pollution.

Future Researchers. Finally, the result of this research will help the future researchers in their own research. It will be a guide to them as they plan and choose their research topic and in working on it.

REFERENCES

Electronic Sources

Adejoke, Olufemi C., Mji, Andile & Mukhola Murembiwa S. (2014).Students' and Teachers' Awareness of and Attitude towards Environmental Pollution: A Multivariate Analysis Using Biographical Variables. Date Retrieved: April 12, 2019.https://www.tandfonline.com/doi/abs/10.1080/09709274.2014.11906690.

Ambag, Rafael(2018). How Bad Is Air Pollution In The Philippines? Retrieved: April 12, 2019. <https://www.flipscience.ph/health/how-bad-air-pollution-philippines/>.

Aygun, M., Ceylan, D. & Gungoren, S., (2017) . Air Pollution Awareness in the Scope of the Community Service Practices Course: An Interdisciplinary Study. Retrieved: April 12, 2019. < https://files.eric.ed.gov/fulltext/EJ1181024.pdf>.

Barnett AG, Knibbs LD. Higher fuel prices are associated with lower air pollution levels. Environ Int. 2014;66:88–91. [PubMed] [Google Scholar]

Becker R, Henderson V. Effects of air quality regulations on polluting industries. J Polit Econ. 2000;108:379–421. [Google Scholar] Stavins RN, Whitehead B. The Greening of America's Taxes: Pollution Charges and Environmental Protection. 1st ed. Cambridge, MA: Center for Science and International Affairs, John F. Kennedy School of Government, Harvard University; 1992. [Google Scholar]

Delgado, Avelino N., (2018). New Special Issue (open to submission, Journal of Environmental Management, JCR Impact Factor 4.010): Technically-based use of by-products as a tool to control pollution. Date Retrieved: April 12, 2019. https://www.journals.elsevier.com/journal-of-environmental-management.

Faiz A, Weaver CS, Walsh MP. Air Pollution from Motor Vehicles: Standards and Technologies for Controlling Emissions. 1st ed. Washington, DC: World Bank Publications; 1996. [Google Scholar]

Goldemberg J, Johansson TB, Reddy AK, Williams RH. Energy for the new millennium. Ambio. 2001;30:330–7. [PubMed] [Google Scholar]

Hall JV, Brajer V, Lurmann FW. Air pollution, health and economic benefits – Lessons from 20 years of analysis. Ecol Econ. 2010;69:2590–7. Google Scholar]

Haines A, Smith KR, Anderson D, Epstein PR, McMichael AJ, Roberts I, et al. Policies for accelerating access to clean energy, improving health, advancing development, and mitigating climate change. Lancet. 2007;370:1264–81. [PubMed] [Google Scholar]

Holman C. Sources of air pollution. In: Holgate ST, Samet JM, Koren HS, Maynard RL, editors. Air Pollution and Health. 1st ed. London: Academic Press; 1999. pp. 115–48. [Google Scholar]

Jimeno, Karen V. (2018). Are Filipinos experiencing 'wildfire' pollution every day? Date Retrieved: April 12, 2019. https://www.philstar.com/business/motoring/2018/11/28/1872203/are-filipinos-experiencing-wildfire-pollution-every-day#zHuwDzBeP71h3Rge.99.

Jacobson MZ. Review of solutions to global warming, air pollution, and energy security. Energy Environ Sci. 2009;2:148–73. [Google Scholar]

Newman NC, Ryan P, Lemasters G, Levin L, Bernstein D, Hershey GK, et al. Traffic-related air pollution exposure in the first year of life and behavioral scores at 7 years of age. Environ Health Perspect. 2013;121:731–6. [PMC free article] [PubMed] [Google Scholar]

Nakano T, Otsuki T. Environmental air pollutants and the risk of cancer. Gan To Kagaku Ryoho. 2013;40:1441–5. [PubMed] [Google Scholar]

O'Keeffe, Jillian (2018). 10 Causes of Air Pollution. https://sciencing.com/10-causes-air-pollution-6890.html>.

Robles, Nathalie, (2018). PH falls third in air pollution deaths. Retrieved: April 12, 2019. < https://www.asianjournal.com/philippines/across-the-islands/ph-falls-third-in-air-pollution-deaths%e2%80%af/>.

Wong EY, Gohlke J, Griffith WC, Farrow S, Faustman EM. Assessing the health benefits of air pollution reduction for children. Environ Health Perspect. 2004;112:226–32. [PMC free article] [PubMed] [Google Scholar]

Tomacruzm, Sofia (2018). Air pollution deaths 3rd highest in PH. Date Retrieved: April 12, 2019. <https://www.rappler.com/nation/208192-air-pollution-deaths-3rd-highest-philippines>.

Williams, Shawna, (2018). Proposed EPA "Transparency" Rule Criticized. Retrieved: April 12, 2019. < https://urbauerlab.uga.edu/fyo/war-2018-html-pdf/Proposed%20EPA%20Transparency%20Rule %20Criticized%20%7C%20The%20Scientist%20Magazine.pdf>.

Other Sources

Philippine Clean Air Act of 1999. Date Retrieved: April 12, 2019. http://www.chanrobles.com/philippinecleanairact.htm#.XLC4HFUzaHs.

The Cost of Air Pollution Strengthening the Economic Case for Action. Date Retrieved: April 12, 2019. <http://documents.worldbank.org/curated/en/781521473177013155/pdf/108141-REVISED-Cost-of-PollutionWebCORRECTEDfile.pdf>.

Indoor Air Pollution. http://eschooltoday.com/pollution/air-pollution/what-is-indoor-air-pollution.html

Effects of Air Pollution. Date Retrieved: April 13, 2019. http://eschooltoday.com/pollution/air-pollution/effects-of-air-pollution.html

Causes of Air Pollution. Date Retrieved: April 13, 2019. https://sciencing.com/10-causes-air-pollution-6890.html

Causes of Air Pollution. Date Retrieved: April 13, 2019. https://phys.org/news/2016-04-air-pollution.html

Solutions of Air Pollution. Date Retrieved: April 13, 2019. https://www.conserve-energy-future.com/causes-effects-solutions-of-air-pollution.php

Air pollution leads to millions of ER visits for asthma attacks worldwide. Date Retrieved: April 13, 2019. https://www.sciencedaily.com/releases/2018/10/181025083955.htm

Environmental Management. Date Retrieved: April 13, 2019. https://www.journals.elsevier.com/journal-of-environmental-management